SAFE ROOM DESIGN

CBRNE Ltd

PUBLISHED BY CBRNE LTD

Title:	Safe Room Design	
Date:	August 30, 2014	
Author(s):	N Hale, D Kelly	CBRNE Ltd

This project from which this manual has been produced has received funding from the European Community's Seventh Framework Programme. The views expressed in this document are purely those of the writer and may not in any circumstances be regarded as stating an official position of the European Community.

Front Cover Design by: Carolyn Smith BA (Hons) Ind Des MFA - CBRNE Ltd Design Director

Glossary of Acronyms

CBRN	Chemical, Biological, Radiological, Nuclear
CCTV	Closed Circuit Television
COLPRO	Collective Protection
CPNI	Centre for the Protection of National Infrastructure (UK)
DNA	Deoxyribonucleic acid
EU	European Union
FEMA	Federal Emergency Management Agency (USA)
FP7	Framework Programme 7
HEPA	High Efficiency Particulate Air (Filters)
IMCOSEC	IMprove the supply chain for COntainer transport and integrated SECurity simultaneously
PRACTICE	Preparedness and Resilience Against CBRN Terrorism using Integrated Concepts and Equipment
TEDA	Triethylenediamine
TRL	Technology Readiness Levels
TV	Television
WC	Water Closet
WiFi	Wireless Fidelity
WP	Work Package (of Project PRACTICE)

Contents

3

1. Executive Summary

This report provides guidance in the specification of Safe Rooms designed as a response to the threat to people from Chemical, Biological or Radiological (CBR) incidents. It does not provide guidance in respect of protection from energetic missiles or explosives. The document is to be considered as a route map for those interested in developing their own Safe Room, not detailed design advice. It provides sufficient detail such that an interested party (e.g. a school, office or factory) can understand the key design parameters that they will need to specify to a detailed design specialist. It also provides a stepwise process (to be undertaken with the help of expert parties e.g. security experts, architects etc) to the development of a detailed specification that can be handed-on to other expert parties to develop the detailed solution. Its use will thus ensure that the user is an "intelligent customer" who procures what is needed rather than what is simply available or offered.

The report examines the key functional requirements to be addressed when designing a Safe Room: these are identified as Isolation, Sustenance, Access and Communications.

Issues to be considered in providing these functional requirements are examined and best practice guidance is presented. Finally a design selection process for Safe Rooms is presented which is driven by an assessment of Design Basis Threat[1]. For non-technical users of Safe Rooms this is considered to be preferable to the simple purchase and installation of off-the-shelf products or products designed to one of the established "classes" such as those defined by the Federal Emergency Management Agency in FEMA 452 for example.

Consideration of the issues associated with the activation of a Safe Room by commercial organisations (rather than military organisations for example) has led to the conclusion that the benefits from inclusion of decontamination and monitoring facilities in a Safe Room are likely to be outweighed by the difficulties in their operation. Furthermore, if such facilities are required prior to allowing personnel into the Safe Room then it is likely that it is already too late to use them. Thus Safe Rooms – as set out in this report – are proposed as part of a planned response to a raised threat or a response to an imminent threat rather than a response to an incident which has already caused personnel contamination. Decontamination facilities are therefore not addressed in this document.

This document can be used in two ways; users can either read through the guidance presented in Sections 3 to 6 and then use the flowcharts in Section 7 to help them generate a specification (this is the preferred route) or they can jump straight into Section 7 and refer back to the guidance as indicated by the pointers on the flowcharts.

---

[1] A "Design Basis" threat is a defined threat or scenario, which encapsulates the scope of threat that is to be designed for – i.e. anything worse than the design basis threat is to considered to be beyond the scope and by definition does not need to be explicitly addressed. The reasons for choosing this boundary can, for example, be based on risk, cost, finance or political factors. It is a concept commonly used in high risk industries (see IAEA).

2.   Introduction

The use of designated refuges or Protected Areas, including specifically equipped Safe Rooms, is a popular strategy for addressing physical threats to the safety of personnel within buildings. Their use provides both a pre-planned and organised response and reassurance whilst providing protection. They are very common as a response to the risk of fire and also to extreme weather events. In both these cases, the nature of the threat is widely understood and experience of real incidents has been used to develop industry practice and official guidance that can be referred to when developing incident management plans and designing Safe Rooms.

The use of Safe Rooms can also, however, form part of a planned response to a terrorist incident or the threat of imminent terrorist attack, particularly where hazardous materials in the form of Chemical, Biological or Radiological (CBR) agents may be used as weapons. The spectrum and nature of CBR agents and delivery methods that might be used in an attack present a range of specific challenges to be addressed when designing Safe Rooms to provide protection against these types of threat. However, Safe Rooms are only one of the possible safeguards against a CBR incident and they should not be viewed as a solution in themselves but as a last line of defence. To make Safe Rooms more effective, or perhaps even unnecessary, the aim should be to remove or at least minimise the threat by adopting effective security measures such as those presented by the United States' Federal Emergency Management Agency (see FEMA 452) and the United Kingdom's Centre for the Protection of National Infrastructure (see CPNI). These measures make an attack less likely by making it much more difficult. Even with best practice security measures in place there may still be a credible threat, or potential for exposure to a release of hazardous material arising elsewhere, for example as a result of an industrial accident, that would require the availability of an effective Safe Room.

This report:

- Identifies the key functional requirements for a Safe Room (Section 3)

- Provides a brief overview of hazardous (CBR) materials of concern and authoritative references to detailed information (Section 4)

- Identifies the key issues in meeting the key functional requirements (Section 5)

- Identifies best practice guidance from official and industry sources relevant to the design and implementation of Safe Rooms (Section 6)

- Presents a process for the design of Safe Rooms based on definition of a Design Basis Threat and assessment of key risks and constraints, drawing on the earlier analysis of key functional requirements and best practice guidance (Section 7).

The authors recommend that users wishing to derive an outline Safe Room specification suitable for their requirements first review and consider the information presented in Sections 3 to 6 and then use the flowcharts in Section 7 to guide them through the process. It is essential that key decisions and the reasons for them are recorded at each stage as evidence of due diligence and to allow the continued validity of the design to be tested.

3.1   Definition of a Safe Room

A Safe Room is a designated and equipped location that may be used as a temporary place of refuge from an external threat, or range of threats. Use of a Safe Room is effectively an enhanced pre-prepared form of "Sheltering in Place"[2], where people are instructed to make use of the natural protection offered by their location or one close by; much of the guidance related to sheltering in place is therefore relevant.

In the context of a CBR threat, the primary function of a Safe Room is to protect building occupants from coming into contact with any hazardous materials present as a result of an incident or attack.

3.2   Top-Level Functional Requirements

The primary functional requirements for a Safe Room are:

- Isolation

    Physical separation to keep the Safe Room free of the Hazardous Material

- Sustenance

    Provision of an environment and essential facilities necessary to sustain a defined number of occupants during their stay in the Safe Room

- Safe Access

    A means of entry into and exit from the Safe Room that does not interfere with its effectiveness (i.e. the safe room should be accessible to those wishing to use it without them being exposed to CBR Agents in the process of getting to it or entering into it).

- Communication

    A safe means of communication to allow occupants to provide information about their status and that of the Safe Room and to obtain information about the outside world, particularly the availability of safe egress routes

Users may also wish to be able to use the Safe Room to provide a base for continued operation of services other than those essential for the operation of the Safe Room itself. This might well affect the choice of Safe Room location and impose other constraints that will need to be taken into account when designing the Safe Room, for example additional waste heat that may need to be allowed for. However, the top-level functional requirements for the Safe Room will still apply and will need to be managed within those constraints. Equipment necessary to provide continued

---

[2] "Shelter-in-place" means to take immediate shelter where you are—at home, work, school, or in between. It may also mean "seal the room;" in other words, take steps to prevent outside air from coming in (see CDC).

operation from within a Safe Room will be specific to a particular business and location and so is beyond the scope of this report.

4.   CBR Threats

4.1   Characteristics of Hazardous Materials

The CBR threat is posed by hazardous materials released deliberately in an attack or through accident. In the case of a deliberate release, it is likely that the material and initial dispersal method will have been selected to maximise the hazard, whereas the range of materials that might be encountered as a result of accidental release is potentially much greater.

4.1.1   CBR Terrorist Weapons

Those materials most likely to be used by terrorists in an attack are both well documented and subject to detailed analysis and continuous review (see for example publically available documents, such as NIOSH 2007, CIA 2003 and Shea 2003).

They are often classified based on their harmful effects and include:

Blister Agents

Chemicals that burn or blister the skin, eyes and soft tissue, also known as vesicants. These can also affect air passages and organs. *Examples: Mustard Gas, Lewisite*

Choking Agents

Chemicals that affect the lungs and cause pulmonary oedema (build-up of fluid in the lungs) which results in asphyxiation. *Examples: Chlorine, Phosgene*

Blood Agents

Chemicals that interfere with the ability of the blood to carry oxygen, resulting in asphyxiation. *Example: Hydrogen Cyanide*

Nerve Agents

Highly toxic chemicals that affect the operation of the nervous system. *Examples: Sarin, Soman, Tabun, VX-Nerve*

Biotoxins

Toxic chemicals produced by living organisms. Some are dangerous in very small concentrations, having effects on the nervous system or specific metabolic processes, such as protein synthesis. *Examples: Ricin, Strychnine*

Biological Infectious Agents

Harmful bacteria or viruses that infect the host and cause illness. *Example: Anthrax*

Radiological Agents

Radiation from radioactive material can cause cell and DNA damage. Different types of radiation present different types of hazards. Generally for incidents other than nuclear attacks, radioactive material is most dangerous when presented in a form that allows it to enter the body, typically through ingestion or inhalation. *Examples: Cobalt-60, Caesium-137, Plutonium.*

4.1.2 Other Hazardous Materials

Also of concern is the wide range of more generally used hazardous materials that could present a risk if released. This includes industrial or agricultural chemicals, biological materials and radioactive materials. Whilst many of these materials are generally less hazardous at low dosages, they are more widely available and are more likely to be stored and transported in larger quantities. *Examples: chlorine, ammonia, hospital wastes and radioactive wastes.*

5.   Meeting Safe Room Requirements

The key requirements identified in Section 3.2 are discussed in subsections 5.1 to 5.4 below. Best practice guidance for each of the key requirements are presented in Sections 6.1 to 6.4 respectively.

5.1   Isolation

5.1.1 Requirement

Provision of protection against the harmful effects of CBR materials is, in principle, very simple. The materials cause harm through physical contact[3] and, in the case of radiological materials only, close proximity. The requirement is therefore to protect Individuals by providing effective barriers between them and any material released into the environment.

5.1.2 Issues

The measures necessary to provide protection from hazardous materials depend on their specific properties in two areas: their physical characteristics (their size and form) and their potential harm mechanism.

Their physical characteristics affect the ways that they may be introduced to, spread and persist within the environment, and so present a risk to any personnel present. This will affect the potential extent and likely duration of the incident.

The potential harm mechanisms depend not only on the physiological effects of the material at a given dose level but also on the limited number of pathways by which the material can enter the body and have an effect. The effective toxicity of a particular material may vary dramatically depending on how it enters the body. A particular material that is extremely harmful if inhaled or ingested, for example, may present relatively little risk of harm through limited skin contact

---

[3] Including direct contact (i.e touch), injection, ingestion and ingestion of vapours or particulate.

(examples includes some forms of radiation which do not penetrate the outer dead skin layers but which are very harmful if ingested or inhaled).

In a known and controlled situation, it may be considered sufficient to provide effective barriers against only those pathways known to be at risk – for example when designing a Safe Room for use against industrial hazards on an industrial site where the range of hazardous materials is already known. In the early stages of a CBR incident, however, it is unlikely that all of the agents potentially present will be identified with sufficient confidence to take this approach. For individuals who have not been contaminated in the initial dispersal, barrier protection, including that provided by the Safe Room, must therefore be designed to deal with the full spectrum of potential CBR agents.

The Source->Exposure->Pathway model illustrates the process of contamination and possible harm as the result of a Hazardous Material release, as shown in Figure 1 below, adapted from PRACTICE WP5 deliverable D5.6 "Protocols for the Justification of Risk from Residual Contamination" (Hale et al, 2012). A Safe Room can be used to interfere with all the pathways shown by isolating occupants from the initial contamination and all the transportation modes (or hazard vectors) shown.

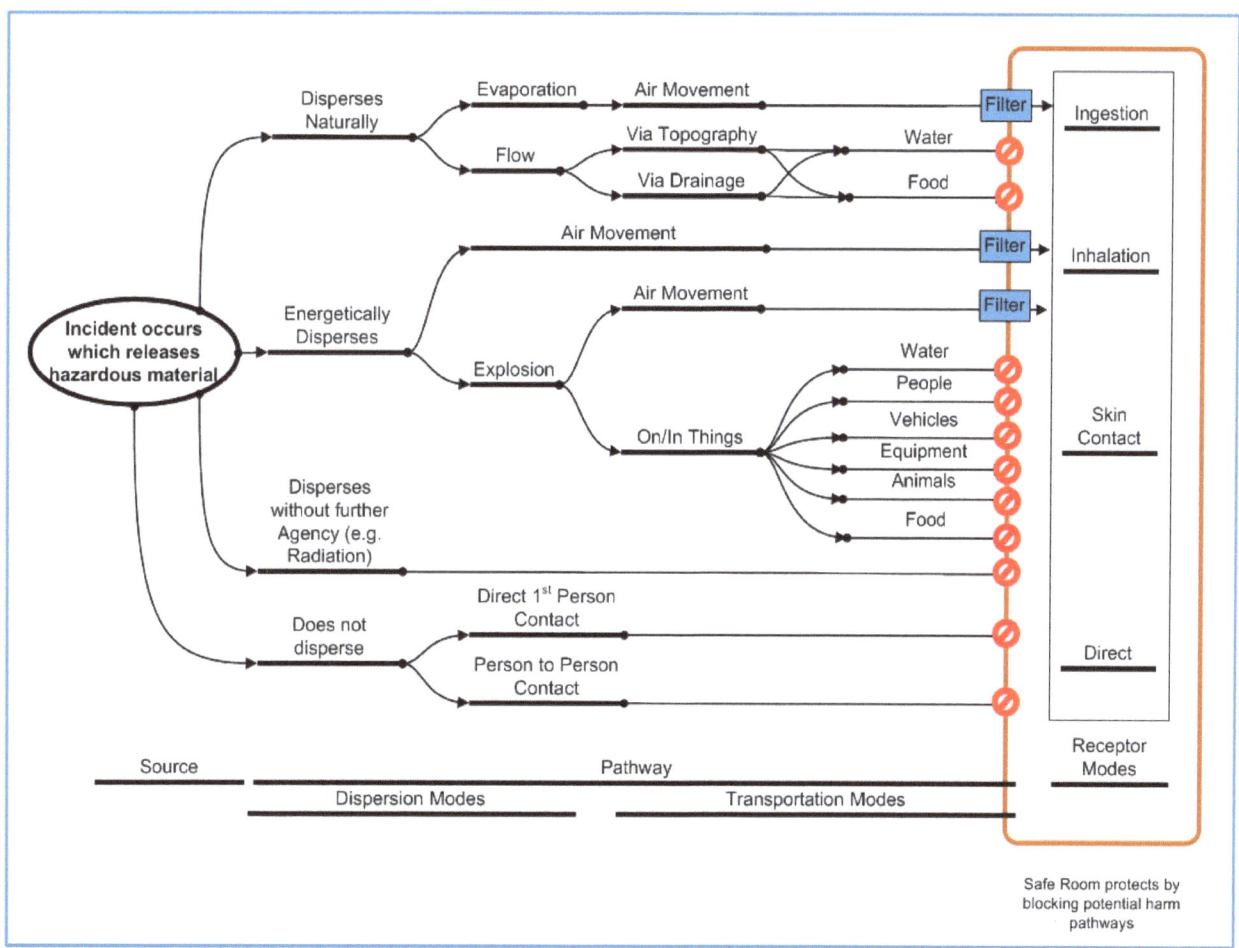

Figure 1: Source-Pathway Model for Safe Room (adapted from Hale et al 2012)

As indicated above, once an incident has occurred, anything that enters the Safe Room is a potential transport medium for hazardous material or, "hazard vector". This may include anyone who may have been exposed to hazardous agents.

The key hazard vectors in the short term are air movement and transport (of contamination) on people and things. The first vector can be addressed by either the continuous provision of uncontaminated air or the provision of a sealed volume of air (thus there are two fundamental types of Safe Room that can be provided, unventilated or ventilated). The second vector is addressed by physically separating the room from the surrounding areas and by not bringing items into the Safe Room – other than people.

If it was necessary to bring anything (including people, equipment and supplies) into the Safe Room from a potentially affected area **after an incident**, effective testing and decontamination procedures would be necessary to exclude the hazardous material and so maintain an effective barrier.

For fast and effective operation of the Safe Room, the need for time-consuming and potentially ineffective testing or active decontamination procedures should be avoided. The most effective way to ensure this is to populate the Safe Room with any required equipment and supplies prior to the risk of any contamination. This requires the Safe Room to be permanently stocked or prepared in advance once the perceived threat reaches an agreed level. This offers the potential to accurately site the Safe Room and include it within a regular maintenance policy. Best practice guidance on achieving the requirement of Isolation is presented in Section 6.1

5.2   Sustenance

5.2.1   Requirement

The requirement is to provide the supplies and facilities in the Safe Room to keep the occupants healthy and reasonably comfortable during their stay regardless of the external hazards from an incident.

5.2.2   Issues

The number of occupants and duration of the stay are the key drivers in determining the amount of equipment and supplies that will be needed in the Safe Room. The most important considerations are air, water, food, light and basic sanitation, because they will be needed first and are required to keep people healthy. Other considerations that may become relevant for longer occupation times are sleeping facilities and entertainment.

In the immediate period after an incident, provision of adequate air is the most important consideration when specifying a Safe Room. The amount of air required to meet a specific quality requirement – in this case breathability – can be determined as that necessary to keep the level of the dominant contaminant within its quality limits. The dominant contaminant is the one that will reach its quality limit first. In a sealed volume of air, human respiration will gradually deplete oxygen levels and increase levels of carbon dioxide (CO_2) and humidity. Of these changes, humans are most sensitive to the increase in carbon dioxide, which becomes progressively more toxic at higher concentrations.

Air supply requirements for Safe Rooms can therefore be determined as those required to keep carbon dioxide levels within safe limits. It may be desirable, however, to also monitor oxygen levels within the safe room where there is a risk that in-leakage of other gases or vapours could lead to a reduction in the oxygen level. The Safe Room should, of course, be designed to avoid this risk wherever possible.

The next most important requirement for sustenance is for drinking water supplies. For longer term occupation food and sanitation become important too.

5.3 Access

5.3.1 Requirement

The requirement is to provide access to the Safe Room whilst recognizing the requirements arising from an incident involving hazardous materials. This is a key requirement that will have a significant effect on the performance and feasibility of a Safe Room.

5.3.2 Issues

There are three issues associated with access, namely prevention of contamination ingress, security and location.

Preventing Contamination Ingress (on things and people)

Preventing contamination of the Safe Room is essential if it is to be effective. Access to the Safe Room by anyone who may have been contaminated will therefore present a problem. An apparently obvious solution to this would be to provide decontamination facilities to make these people safe to enter the Safe Room. Decontamination airlocks are a common feature of many collective protection (COLPRO) facilities for users with high levels of technical competence and resources and guidance on their design and use are readily available (US-ACE).

However, decontamination airlocks may not be appropriate for an emergency Safe Room for non technical users for several reasons:

- Effective decontamination needs to be done carefully and methodically and requires training. If staff entering the Safe Room are hurried or untrained, there will be a significant risk that decontamination will not be wholly effective, potentially exposing everyone in the Safe Room. Demonstration of effective decontamination (and determination of appropriate techniques) also requires that detection / monitoring equipment is available and that operatives are trained in its use.

- External decontamination is not effective against material that has already entered the body. It will be impossible to decontaminate personnel exposed to an infectious biological agent to prevent them becoming a risk to others.

- Decontamination airlocks require a significant flow of air to expel any material released into the air during the decontamination process, so will not be suitable for unventilated Safe Rooms.

- The time required to decontaminate and monitor persons needing to enter the room may conflict with the need to get people to safety very quickly to protect them from an imminent threat or spread of contamination.

The guidance developed in Section 7 of this report assumes that decontamination facilities and associated airlocks are not to be included in the design. This is of course a matter for the user to confirm when developing the Statement of Requirement and concept of operations for the Safe Room.

Providing security

The entrances to the Safe Room may need to be secure against forced entry. As well as protection from any hostile act targeting the Safe Room, this is to prevent people who may be contaminated and trapped outside the Safe Room after it has been sealed from gaining entry and thereby risking the safety of those already in the Safe Room.

The problem of what to do about people who may either already be contaminated, or who are not able to get to the Safe Room whilst it is safe for it to remain open, is significant and will need to be addressed. This aspect provides an argument in favour of having multiple Safe Rooms (or at least separate compartments) which can be filled and sealed in turn, allowing those who have already reached the Safe Room to be afforded maximum protection, but still offering protection for new arrivals.

Where the area outside the Safe Rooms is contaminated, continued exposure will lead to increased dose received by those left outside, so access to a separate Safe Room from those who have not been contaminated will be beneficial even for those who have already been exposed. Accommodating people who have been exposed separately prevents them from becoming a hazard vector and a risk to others. This is most important for hazardous materials dispersed as liquids, droplets or fine powders because they are readily transportable once they come into contact with an individual.

A much lower risk is presented by some materials, particularly vapours or gasses that are much less likely to be transported in harmful quantities by exposed individuals. However, because it will not be possible to immediately identify the hazardous material (and so assess the risk) the precautionary approach of strict separation between exposed (or suspected exposed) and not exposed individuals is recommended.

Location

If the Safe Room is to be used in response to an incident it is vital that it can be activated, populated and sealed within a suitably short period. How short that period needs to be will be determined by an assessment of the specific types of incident for which the best response is sheltering in the Safe Room. Close proximity of the Safe Room to the people it serves will help to reduce the time before the Safe Room can be sealed and thus provide protection[4].

Guidance and metrics used in Fire Safety Risk Assessments, specifically the study of routes and maximum travel distances for evacuation in case of fire may also be used to evaluate Safe Room

---

[4] This also offers the possibility of segregation/selection of personnel depending on what part of the building they were in at the time of the alarm; large covered communal working spaces offer a higher risk than the confinement of tight corridors with numerous doors offering natural protective barriers.

locations if the Safe Room is treated as an evacuation point. Egress routes out of the building from the Safe Room can be similarly evaluated.

5.4 Communication

5.4.1 Requirement

The two most important communication requirements for a Safe Room are to inform the emergency services that the Safe Room has been activated, including the number and condition of occupants, and for the occupants of the Safe Room to obtain information about the progress of the incident and instructions from responders, particularly regarding the availability of safe egress.

5.4.2 Issues

In normal circumstances, a wide range of communications systems may be in use or available. However, it is important to consider whether they will continue to be available during or after an incident or threat. Public communications infrastructure could be rendered inoperable as a direct result of an incident, or it could be deliberately turned off (e.g. on instruction of the emergency services) or otherwise made unavailable for normal use in response to the incident. Where mobile telephone and messaging systems continue to be available, they could easily be overwhelmed as people turn to them in response to an incident. They should not, therefore, be relied upon as the sole means of communication from within the Safe Room. Wired "landline" telephones and wired networks are likely to be more reliable as long as the physical infrastructure is not damaged, though again there will be potential for public systems to become overwhelmed with traffic.

In all cases, it is important to establish and test emergency communications links in advance.

6. Best Practice Guidance

The following key points on construction and operation of Safe Rooms are taken from current guidance from UK, European and US official sources (see the numerous documents listed in the References and Further Information sections)

Sections 6.1 to 6.4 specifically address the key functional requirements discussed in Section 5 while section 6.5 addresses other important considerations.

6.1 Isolation

Most published guidance concentrates appropriately on Air Movement as the primary hazard vector, particularly into and within buildings.

The primary concern is "infiltration" or uncontrolled airflows into the Safe Room from the outside environment, which occur whenever there is a local pressure differential such that the pressure at a leakage point is higher outside the room than inside it (see Figure 2 for an explanation of the terms infiltration and exfiltration).

This concern can be addressed by sealing leakage points to make a "tight" structure. Whilst a hermetically-sealed room will provide protection, it is often very difficult to achieve in practice, so it is difficult to eliminate infiltration completely. Studies have shown that even with very limited levels of infiltration from a contaminated environment, contamination will slowly build-up within the Safe

Room and will be retained longer than a typical exterior plume of contamination would take to disperse. As the outside contamination disperses, or contamination builds up and is retained in the Safe Room eventually the level of contamination within the Safe Room would be higher than that outside and the Safe Room would no longer be providing any protection. This is one of the main arguments put forward against the use of simply sealing a room and "Sheltering in Place", however for short durations where the outside contamination disperses rapidly, an unventilated Safe Room can provide effective protection (MEFB 2011). There is an obvious limitation in the length of time that any sealed room can be used, because of the limited supply of breathable air (this is discussed further in Section 6.2 "Sustenance").

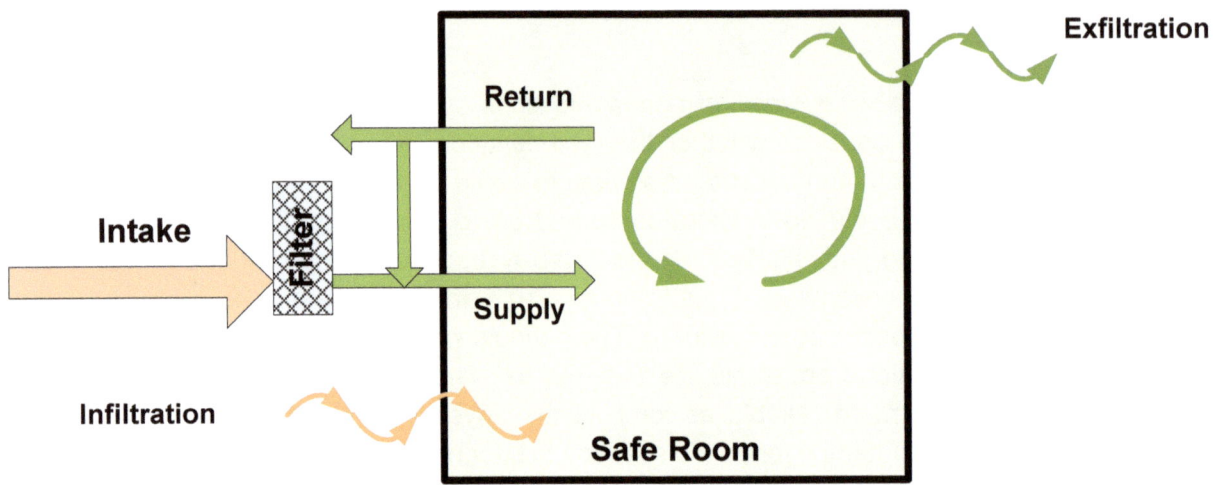

Figure 2 Schematic of a (Ventilated) Safe Room Showing Airflows

The problem of infiltration due to inadequate sealing can be addressed by ensuring that pressure differentials at all leakage points are positive – i.e. that the pressure inside the room is higher than that outside, so that all air movement will be from the room to the outside. The obvious implication of this is that the air inside the room will need to be replenished to maintain the required pressure differential. For typical Safe Room applications, the quantities of air required are too large for the use of stored air to be practicable, so it becomes necessary to use air from outside the safe room, which must be made safe by effective filtration. Replenishment of the air in the Safe Room with sufficient outside air enables the Safe Room to be used for much longer periods if sufficient supplies and filter capacity are available. As a breathable air supply may be required regardless of the filtration requirements, air supply and filtration is discussed further in Section 6.2 "Sustenance").

6.1.1 Sealing the Safe Room

Simple measures, like taping over gaps around windows, doors and utility points (power, water etc) or covering them with polythene sheeting tightly taped around the edges have proven to be effective, when done carefully. Much of the basic guidance for Sheltering in Place intended for the general population has included this approach.

One major drawback of this "Expedient Sheltering in Place" is the significant time it takes to tape and seal a room effectively – leading to vulnerability in the early stages of the incident and hurried work that can result in ineffective sealing (Sorensen J H, Vogt B M 2001). Time to apply seals can be reduced by pre-cutting some of the materials used and rehearsing, or at least pre-planning, the sealing process. [5]

The most effective approach is, therefore, to seal the safe room as much as possible in advance, only leaving access doors and any vents designed to be in operation during an incident unsealed.

All surfaces and features of the Safe Room should be examined to confirm that they are air-tight unless they need to be otherwise. Non-obvious areas that may require extra sealing include false ceilings or floors, ducting for services including power, lighting, water and sewerage and any hollow pipes or cables, open joints between wall and/or floor panels that may lead to shared cavities. There may also be a degree of leakage through gas-porous construction materials (e.g. some concrete, brickwork, timber) rather than at joints between them, requiring the use of sealant coatings (CMHC 1998).

Smoke seals on fire doors can provide some degree of protection and will be in place prior to any incident. They are designed to limit spread of smoke at ambient temperatures unlike intumescent seals designed to provide enhanced sealing in response to elevated temperature (i.e. intumescent seals do not provide effective seals during normal temperature conditions). They may provide an appropriate level of sealing for a ventilated safe room where a controlled level of leakage is required for a positive pressure system, or may provide limited protection whilst more effective temporary sealing is applied.

The construction methods and materials used in older buildings tend to be more "leaky" than recent buildings, making it more difficult to seal them well (MEFB 2011).

A ventilated Safe Room cannot, of course, be completely sealed, though it will be preferable to seal uncontrolled leak points wherever possible so that overpressure and the flow pattern through the Safe Room can be regulated by controlled venting.

6.1.2 Preventing Infiltration

The actual overpressure required for an effective barrier to infiltration depends on a number of factors, particularly the incident (wind) pressure to be overcome and the shape and orientation of the space. When considering infiltration in a whole building, for example, the number of floors has a significant effect. Overpressure requirements to prevent infiltration into a whole building are likely to be greater than overpressure requirements to prevent infiltration into the Safe Room from elsewhere in the building. Whilst the required overpressure for a particular structure can be modelled and calculated, practical tests will be required to confirm that it is effective in preventing infiltration.

The main limiting factor in achieving an appropriate positive pressure is the "tightness" of the area to be pressurized. The greater the amount of air that can leak out, the greater is the amount of air that must be pulled in and filtered. The most common method for maintaining a specified

---

[5] Heavy duty tailored impermeable material and an industrial standard Velcro sealing system around the area that requires sealing has previously been used and tested. This method can be used to partition off rooms and build an effective controlled area / multiple safe room locations within a corridor.

overpressure is to fit a pressure controlled vent that will open at the specified pressure differential. Of course this requires sufficient airflow to create the positive pressure despite the presence of any leaks.

If uncontrolled air leakages out of the Safe Room (exfiltration) can be largely eliminated then it becomes possible to pressurise a larger area. In some cases it may be feasible for the Protected Area / Safe Room to be expanded to cover a whole floor, for example, or even the whole building. If the whole building approach is taken it is still important to compartmentalise the air handling systems to avoid exposing the entire protected area to any internal release.

One significant advantage of the whole building pressurization approach is that infiltration into the building is potentially completely preventable in the case of an external release. Time consuming and expensive internal decontamination can thus be avoided and the building can be brought back into use much more quickly, reducing the impact and "value" of the incident to any perpetrator.

6.2   Sustenance

The fundamental requirement for sustenance is breathable air and therefore this is the focus of much published guidance.

For a short stay of 1 to 2 hours the additional requirements are only for very basic facilities and minimal supplies. For durations of between 2 and 12 hours the list should be expanded to include food, seating and entertainment (such as TV, reading materials and non-energetic games) to help pass the time.

For stays of over 12 hours basic sleeping facilities such as cots or mats should be considered.

6.2.1 Air

Air Quality

Normal atmospheric air contains approximately 0.038% carbon dioxide by volume (Conway/NOAA 2011). The safe limit for continuous exposure in a normal working environment is approximately 1%. Carbon dioxide begins to have significant physiological effects - leading to heavy breathing and loss of cognitive ability - at about 3%, which is considered the maximum safe limit for short term exposure. At 15% concentration CO_2 is likely to be fatal within minutes, though there is a risk of fatalities at concentrations as low as 7%. (Harper 2011)

The suggested upper limit for carbon dioxide within a safe room, for design purposes and as a trigger value requiring immediate ventilation or evacuation of the safe room is therefore 3%. All Sealed Safe Rooms and Safe Rooms where the ventilation is close to the minimum must be fitted with a continuous CO_2 monitor with an alarm level set no higher than 3% CO_2.

Respiration studies indicate an approximate mean rate of CO_2 production of approximately 0.5 of a litre per minute per person for average adults performing limited activity (NIOSH/Foster-Miller 2007). Based on this, starting from a nominal CO_2 concentration of 0.038% the minimum initial free air volume needed for CO_2 levels to remain below 3% in a sealed safe room is approximately one cubic metre per occupant per hour of required occupancy, and approximately three cubic metres per occupant per hour to remain below the 1% long term exposure limit.

Similarly, to keep CO_2 levels below 3% by continuous ventilation in a vented room, the required minimum air replacement rate is approximately 17 litres per person per minute, and over 51 litres per person per minute to remain below the recommended continuous exposure limit of 1%.

These are absolute minimum levels and an appropriate safety factor must be applied to any Safe Room specification to account for items such as the following:

- the level of activity might be higher than that in the studies (largely sedentary with occasional moderate activity and no significant physical exertion)

- the maximum duration may not be clearly defined

- the maximum number of occupants may not be clearly defined (e.g. visitors who may be in the building at the time of the emergency)

- the occupants may include people with breathing difficulties or a medical condition that could make them more vulnerable to the physical stress imposed by an abnormal atmosphere.

- Air quality (temperature, humidity, odour) considerations may require a greater capacity.

The above figures assume the ideal conditions of perfect mixing of the air in the Sealed Safe Room, giving an even distribution of CO_2 and first-in-first-out flow straight through the ventilated Safe Room. For ventilated Safe Rooms, the air flow pattern through the Safe Room should be designed to avoid any still areas. For unventilated Safe Rooms, fans will help to mix the air and prevent local build-up of CO_2. For unventilated spaces, studies have shown that fans can significantly increase occupant comfort and the apparent freshness of the air, particularly in warm or hot conditions (Moran et al) .

Air Intakes

Outside air intakes should be located to satisfy two main requirements:

- They should be located in the area least likely to be affected by a release or an accumulation of agent. They should therefore be high up and not in a sheltered area where still air might accumulate.

- They should be located in an area that is not easily accessible, to prevent them being interfered with or used as a potential delivery mechanism for a CBR agent.

Further guidance on air intake location and construction and other ways to reduce vulnerabilities can be found in FEMA 426.

Clean Air Ducting

Inducted air can be carried to the Safe Room using normal ventilation ducting although segregated and protected dedicated ducting will provide optimal protection and may be easier to maintain.

It may be necessary to filter the air supply (to remove hazardous agents) as discussed below and the choice of filtration and fan location will affect the ducting integrity required.

Where the air is filtered only at the air intakes, measures should be taken to ensure that the air is not contaminated before it reaches the Safe Room. Either the ducting should be completely sealed and tested to prove it is so, or the ducting should operate under positive pressure to ensure that any leakage is away from and not into the air supply duct. To achieve positive pressure in the duct it will be necessary for the air to be pushed using fans near the intake, rather than pulled using fans in the Safe Room. This may require remote control of the fan operation from the Safe Room.

The Safe Room must be isolated from other areas, so that the intake ducting between the filters and the Safe Room cannot be shared, other than with another Safe Room. Safe Rooms linked to a common air supply in this way would have to be managed and operated as though they were a single Safe Room. Outflow ducting should be fitted with a pressure-operated non-return device to prevent infiltration through the ducting when the Safe Room is not operating at positive pressure.

Filtration

No single filter type is capable of removing all potential agents from air. It is necessary to use a cascade of filters, each filter dealing with agents with specific physical or chemical properties (see Table 1 for examples).

The types of filtration needed for a particular Safe Room installation can be determined at the design stage based on an assessment of the threat, the cost of the filter installation and replacement filters and the overall risk assessment (NIOSH 2003).

During any incident or heightened threat scenario, it is very unlikely that the agent / contaminant involved can be identified early enough to permit selection of suitable filter types. If an incident occurs then the most effective filtration available should be used. The selection and installation of filters is a highly technical subject and will require expert guidance.

Table 1 Air Filter Types for Hazardous Materials

Filter Type	Used For	Effective against	Issues	Cost / Availability
Large particle / rough filter	As a first level of filtration on outside air intakes, trapping large particles and reducing the amount of debris that reaches the higher density filters	Large particles, insects		Low cost, widely available
Prefilter / standard HVAC filter	A low-cost, less effective pre-screening filter used upstream of the HEPA filter to prolong its life.	Smaller particles down to approx. 3 microns, larger airborne dust particles and pollen / spores		Low cost, widely available.
HEPA – High efficiency particle arrestor / High Efficiency Particulate Air	As a high efficiency filter to remove the smallest solid particles from the air.	Airborne solid particles and aerosols, typically rated at 99.97% effective at removing particles larger than 0.3 microns. Pollen / spores / viruses E.g. Anthrax spores at 1 to 5 micron, Bacteria 0.3 to 1 micron	Will cause a significant pressure drop across the filter, particularly when the filter is dirty / loaded. This will require higher power fans, leading to higher energy consumption.	Moderate cost, widely available
Adsorbent filter – a filter with a large active effective surface area to which contaminants will bind, typically based on activated carbon.	To capture most chemicals by adsorption onto the filter material. Can be made more effective against selected compounds.	Most Chemical Weapon agents and many industrial chemicals. Simple activated carbon filters may not be effective against gasses and chemicals with high vapour pressure, so military specification filters are impregnated with reactive materials designed to react with the agent to provide effective filtration. One example is ASZM-TEDA carbon, impregnated with copper, zinc, silver, and molybdenum compounds and triethylenediamine (TEDA), as used in military specification filters.	The air needs to be in contact with the filter material for an appropriate time for it to be effective, limiting maximum flow rate. Filter efficiency is reduced at high humidity, so it may be necessary to condition air before filtering, requiring dedicated air conditioning for any dedicated air intakes.	Expensive, special purpose equipment

Morrison (2001) provides a good assessment of the performance of combined HEPA / Adsorption filters challenged with CBR agents.

It is important to recognise that all filtration systems are fractional, i.e., their performance is measured by their efficiency in removing contaminants from the air and unless the filter is rated at 100% efficiency some material will get through the filter. For example, a HEPA-rated filter with a quoted efficiency of 99.97% may allow up to 0.03% of particles through. Filtration effectively operates to reduce the concentration of any contaminant to a level at which it is either harmless, or at which it poses a negligible risk.

Of further concern is that even military-specification adsorption filters cannot protect against certain chemical hazards. Only limited protection is provided against certain acidic gasses, including fluorine, chlorine, hydrogen chloride, sulphur dioxide and others where the filter can be prone to off-gassing (re-releasing) of previously captured material when loaded and used for an extended period. Filtration performance can also be poor at high relative humidity. No protection is provided against certain gasses with high vapour pressure, including ammonia, carbon monoxide, nitric oxide and others.

The performance characteristics of available filters should be understood and taken into account when assessing threats and risks during the design and specification of a Safe Room. Where filtration performance cannot be relied upon, given the spectrum of likely threats, unventilated operation may be preferred, accepting the time limitations imposed by the build-up of CO_2 (see Section 6.1.2) and the requirement for effective sealing (see 6.1.1 and 6.1.2) rather than accepting the risk that some hazardous materials might get through the filters.

Air quality monitoring

The air quality in an inadequately ventilated Safe Room may deteriorate to the extent that it becomes dangerous to remain there regardless of any external hazardous agent. Where a Safe Room may be operated in Sealed/Unventilated mode, it will be important to monitor the air quality to guard against the build-up of toxins, particularly Carbon Dioxide. If the CO_2 concentration in the Safe Room reaches 3%, then either it must be ventilated or evacuated. Equipment to provide continuous monitoring of CO_2 levels is commercially available. It is also advisable to monitor Oxygen (O_2) content of the atmosphere to protect against other oxygen depleting mechanisms. Commercially available equipment is similarly available for this.

Heating & Cooling

The environmental requirements of a Safe Room will be affected by the local climate and weather. If cooling is required as a health issue in normal operation elsewhere in the building, it will also be required for the Safe Room. This can be achieved using air flow, conditioned or cooled as necessary for ventilated Safe Rooms. Sealed Safe Rooms will require cooling systems with a heat sink located outside the room. The requirement for cooling will be linked to the required occupancy time.

6.2.2 Water

Provision of drinking water in the Safe Room is very important, since after only a short time, people can begin to dehydrate. It is important that people drink a normal amount of water during the period that they are in the Safe Room. It is not recommended that drinking water is rationed.

Carbonated beverages should be avoided for any Safe Rooms, because they would unnecessarily increase the level of Carbon Dioxide in the Safe Room.

To avoid any risk of contamination to piped water supplies due to a CBR incident, water supplies for use in the Safe Room should be in sealed containers (e.g. bottled) and stored there to be on hand when needed. The supply should be monitored and replaced as required to ensure that it is safe to use when needed. A minimum of 2.5 to 3 litres per person per day should be provided (WHO 2005).

6.2.3 Food

The need for food will depend very much on the length of time that the safe room is likely to be occupied for and although an important consideration it is secondary to drinking water. Emergency food rations should be in sealed containers and stored within the Safe Room, monitored and regularly replaced as required. Consideration should be given to the provision of heating devices like microwaves that do not consume oxygen or produce CO_2 – these may be especially important where mothers with infants are present.

6.2.4 Sanitation

Access to toilet facilities is essential, although very basic facilities may prove to be adequate for short stays with small numbers of people. A key decision will be whether to use existing facilities, reliant upon piped water supplies, or to use self-contained facilities.

Issues to consider:

- Can the piped water supply be relied upon to be both available and uncontaminated in the case of an incident?

- Can any waste traps on sinks, WCs etc be relied upon to maintain an effective seal against air infiltration? (note that pressure variations across the waste traps due to overpressure or ventilation changes could force the water out of the trap and thus break the seal if the variations are large enough. Also, water trap seals may be broken if the water in the traps is allowed to evaporate)

In many cases, the safest solution will be to use portable chemical toilets, in modesty tents if necessary. This is also likely to be more cost effective where conventional facilities do not already exist within the candidate Safe Room area.

6.2.5 Lighting

There are no special requirements for lighting levels in a Safe Room over and above those that are relevant to normal office or accommodation spaces, other than they should not consume oxygen. Electrically powered lighting (mains or battery) is preferable. If Safe Rooms are to serve on-going operational functions then the normal ergonomic lighting requirements for those functions should be addressed.

6.2.6 Comfort and Entertainment

The requirement for seating or sleeping accommodation will depend on the required duration of confinement in the Safe Room. For short stays of up to an hour or so, seating may not be required for all occupants. For longer stays, seating or lying accommodation should be provided for all occupants. At a basic level this could consist of floor mats. Seating and lying accommodation take up progressively more space than standing, so the facilities provided may affect the floor area required.

Because the occupants of the Safe Room may be together in close proximity and in a stressful situation, it is important to provide appropriate means of passing the time. Non-physical games or entertainments such as TV / movies may help to maintain calm. These may be especially important if children are present.

6.3 Access and Location

The safe room location should be specified or chosen based on the following principles:

- It should be close enough for people to get to it quickly in the case of an emergency. For large buildings, multiple safe rooms may be preferable.

- Personnel should not have to cross potentially contaminated areas to get to the Safe Room, so vulnerable areas such as entrance halls or mail rooms should not be relied upon as routes to the Safe Room. Consequently, the Safe Room entrance should not share common ventilation space with any of these vulnerable areas. Compartmentalisation of a building, as may be required by Building Regulations for Fire purposes, provides a means of limiting the impact and spread of a localized event. Measures taken to limit the spread of fire and smoke can be similarly effective at limiting the spread of CBR agents.

- Because some agents are heavier than air - and so may accumulate in low-lying areas - the Safe Room must not be located in a basement and ground-floor locations should be avoided, if possible. This advice may conflict with accepted guidance on location of Safe Rooms designed for use in some other extreme events (e.g. earthquake, weather), where the basement or lower floors are preferred. Because of this it may not be possible to select a single Safe Room location that will optimise protection against both CBR and extreme weather threats.

- If protection against physical attacks, including blast, firearms and other projectiles, including vehicles, is required then among other safeguards, the Safe Room should be located in a structurally sound and strong part of the building, surrounded by structural walls, rather than partitioning. For the same reason, the Safe Room should ideally have no windows and must have none on the outer face of the building.

Multiple smaller Safe Rooms have several advantages:

- Reduced risk of cross-contamination from individuals who are unknowingly contaminated.

- Each Safe Room can be smaller, and so may provide more location options and be easier to equip.

- Having multiple Safe Rooms provides greater resilience by providing more options if one or more Safe Rooms or associated access routes are contaminated or hazardous.

- A single Safe Room could provide an attractive target for a secondary attack. Multiple separate locations, each with smaller numbers of people would be more difficult to attack and less attractive to a terrorist.

- For smaller Safe Rooms, it may be easier to obtain suitable equipment off the shelf rather than having to design a larger bespoke installation.

The principal disadvantages are:

- Multiple safe rooms will require more administration and maintenance.

- Multiple locations will tend to complicate incident management procedures – particularly in establishing communications between multiple locations and co-ordinating headcounts etc.

- Costs may be higher than for larger installations due to loss of economies of scale.

The extreme example of ventilated multiple Safe Rooms is provided by off-the-shelf miniature shelters – effectively filtered and pressurised tents designed to provide a degree of protection from airborne agents. Whilst these provide more protection than simple sheltering in place, they do not provide the physical protection provided by a specially constructed Safe Room, though they may provide an alternative strategy to sealing the room itself and a low-cost route to filtered ventilated positive pressure protection. The enclosures are typically clear-sided providing the illusion of more space when located in a larger room.

Similarly, multiple unventilated Safe Rooms may be provided by allowing for multiple sealed rooms off an existing corridor.

Decontamination

Decontamination points at entrances to the Safe Room are not recommended in most cases; because of the potential for ineffective decontamination or doing additional harm by using inappropriate methods for certain substances - some authorities recommend that all decontamination is carried out by emergency services response teams. In any event decontamination is rarely needed if personnel were under cover (within the same building) and the incident was external to the building. Consideration could be given to providing facilities for disrobing the outer layer of clothing and donning simple disposable suits as this would offer an additional reassurance that everything that could be done, has been done. The practicality of this will depend on available space, numbers of people and the time delays that will be introduced.

There are also issues associated with the requirement for specialist training in the use of monitoring and decontamination equipment. It is necessary to determine if decontamination will be appropriate and effective for the particular contaminant in terms of the impact upon time to enter the safe room. Such a decision will indicate if such facilities are appropriate for Safe Rooms that may be used by civilians or staff who have not been specially trained.

Where the concept of operations for the Safe Room includes rapid occupation by personnel who have not been exposed and where the Safe Room is to be breached and evacuated only once the "all-clear" is given, decontamination as the Safe Room is filled is unnecessary. Decontamination to provide access once the Safe Room has been sealed would present a significant risk to the integrity of the Safe Room.

6.4   Communication

Since any significant incident is likely to attract the attention of news broadcasters, television and broadcast radio may be a useful source of information regarding the incident and the situation outside the Safe Room. Broadcast media are also often used by emergency services to provide direction and safety advice to the public during major incidents. It is important to establish in advance which broadcast sources are likely to provide information and have equipment ready tuned or with pre-sets. Battery powered equipment may be preferable unless a secure source of external power can be provided.

Whilst the emphasis of communications is likely to be finding out about the incident and communicating with emergency services, it is also important for the occupants to know what is happening immediately outside the Safe Room, particularly around the entrance and possible egress routes. CCTV systems may be used to allow monitoring of the surroundings from within the Safe Room, though appropriately strengthened small windows or simple door viewers may provide a simple and robust alternative. A simple intercom or door entry phone should also be installed at the Safe Room entrance to allow communication without breeching the sealed entrance.

The operating procedures for the Safe Room should include details of external emergency contacts and sources of information, including any telephone numbers to call. These details should be permanently fixed in the Safe Room and periodically checked to make sure that the information is current.

It is recommended that multiple communication options including access to broadcast media are provided. In addition, there should be clear instructions within the Safe Room that communicate the procedures and arrangements to follow whilst in the Safe Room.

The construction and location of the Safe Room, away from the outside of the building, may impede radio signals, limiting mobile telephone, wireless networking and TV and broadcast radio signals. Wherever these are expected to be used within the Safe Room, performance should be tested with the Safe Room closed and configured as it would be for an incident. It may be necessary to provide external antennae or (where permitted) signal boosters to provide adequate and reliable performance. Access to broadcast information sources should be a factor in determining the best location for a Safe Room (e.g. radio and telephone signals, Wi-Fi or wired broadband access).

The emergency services should be informed of the existence of the Safe Room when it is first commissioned and not only when it is first activated in response to an incident. This will allow communications links to be set up and also ensure that the emergency services are aware of the Safe Room as an asset that they can instruct the occupant(s) to use it as necessary during an incident (or alert) as an alternative to evacuation (which may be hazardous). The preferred and fall-back means of communication with them should be agreed and should be tested regularly.

6.5 Other Considerations

6.5.1 Safe Room Equipment

Equipment in the Safe Room is provided for the Safety, Sustenance and Comfort of the occupants.

Power

In many cases, it is likely that mains electrical power will still be available, unless it is shut down as part of an emergency protocol. Where electrical power is shut down, it will be necessary to provide alternative power for essential equipment in the Safe Room and any powered ventilation.

For equipment with relatively low power requirements, such as lighting and communications, it may be possible to use battery powered or manually chargeable devices.

Some small in-room ventilation systems can be hand operated as an emergency backup.

Battery systems kept within the Safe Room should be of the sealed type and free of emissions during charging, storage and use.

Equipment selected for the safe room should neither emit pollutants into the air nor use up oxygen. This is essential if the Safe Room may be operated in Sealed Mode.

Safety Equipment

The following safety-related equipment should be provided in the Safe Room.

- Personnel register – a register of personnel expected to be in the Safe Room and a means to record those that are in the Safe Room. This will allow a tally to be passed on to the emergency services and also allow the Safe Room to be sealed at the earliest opportunity once its intended occupants are accounted for.

- Emergency Lighting – necessary if normal lighting relies on mains power.

- Communications – see Section 6.4

- Air Quality Monitors (CO_2 / O_2) – see Section 6.1.2 "Air quality monitoring"

- Basic First Aid kits – this should be monitored and replenished as necessary to keep perishable items in-date.

- Fire Extinguishers - should be provided in numbers as would be appropriate for the space in normal circumstances and to deal with the hazards present. Extinguishers should be of types suitable for use in enclosed spaces. Carbon dioxide, halon and dry powder are all generally unsuitable, as would any other types that would normally require the area to be ventilated before it could be occupied. Water or foam extinguishers are likely to be more suitable.

A typical checklist of Safe Room equipment is provided in Annex 1.

6.5.2 Activation

The Safe Room is most effective and easiest to use in situations where people have not already been exposed to a hazardous substance, so it must be occupied as soon as reasonably practicable after an incident becomes known or as soon as the potential for an incident rises above a certain threshold in situations where use of the Safe Room, rather than evacuation, is determined to be the best course of action in response to the incident[6],[7] (see Figure 3).

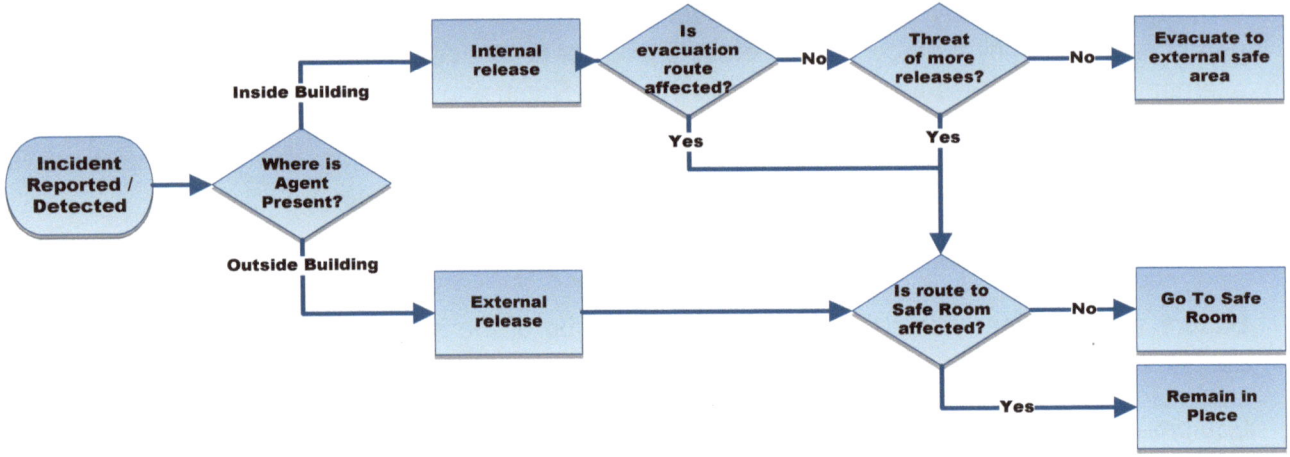

Figure 3 Example Activation Decision Process for Safe Room

6.5.3 Safe Room Operation

People for whom the Safe Room is intended to provide protection should be given clear guidelines for how and when the room may be activated and who has the authority/responsibility to initiate such activation. Similarly, authority/responsibility for declaring the room safe to exit should be clearly defined.

For the Safe Room to provide protection, it is vital that it is occupied and sealed before the hazardous material reaches it, the designated occupants or any part of their route to it. If the Safe Room is to be occupied in response to an incident, this means that the room must be filled rapidly and sealed once all allocated people are present (as registered by the person appointed to be in charge of the Safe Room during the incident), or after a specific short amount of time. The Safe

---

[6] If an alarm is used it must be different to that used in the event of a fire as a fire drill will move people to the outside of the building. The alarm must also satisfy the same requirements for audibility at all locations as a fire alarm.

[7] If a Safe Room were to be activated after people had become exposed to a hazardous substance then its safe use would depend on there being a method of checking those wishing to use it for contamination and for either decontaminating them or segregating them from others.   Some of the issues associated with this have been raised in earlier sections of this document.   Operation and construction of such facilities is outside of the scope of this document.

Room should not be opened again until the "all-clear" is given unless conditions within the Safe Room change and are no longer deemed safe.

6.5.4 Integration with Emergency Planning

The existence and potential operation of any Safe Rooms should be integrated into the fire, emergency and Incident Response planning process for the building. As such, it is likely to be most effective if the operation of the Safe Room, including marshalling personnel, is carried out by the same personnel responsible for evacuation in case of fire, or other incident. In the same way that fire drills are routinely undertaken it is appropriate to also undertake routine drills involving the activation of any Safe Rooms.

If the Incident Response Plan includes an internal assembly point, and the Safe Room is a suitable location with appropriate access and egress, then the Safe Room should be used as the assembly point.

6.5.5 Exposed Personnel / Contaminated Areas

To reduce any risk of spreading contamination, current guidance is that affected personnel should be assessed and decontaminated on-site rather than dispersed. Particularly in the case of Bacteriological and Radiological agents, contamination may not be evident for a long period and the agents are persistent, increasing the likelihood of cross-contamination if people are dispersed before decontamination and testing (HMG 2006).

For this reason, a clear emergency response plan is required, so that people know (or can be instructed) what to do and where to go if they are caught up in an incident. Containing people and any contamination they may carry is important for their safety and that of others.

6.5.6 Safe Room Maintenance

As for other emergency equipment, like fire alarms and extinguishers, a maintenance and inspection plan should be developed for all equipment that composes the Safe Room and its support services. Inspections should include checks that the Safe Room remains accessible and functional given other changes of use and modifications that may have occurred elsewhere on the site. Where practicable these maintenance requirements should be integrated into existing maintenance practices for the site concerned.

7. Specifying a Safe Room to Serve a Specific Building

The United States Federal Emergency Management Agency (FEMA) has produced guidance (see FEMA 453 chapter 3) that classifies Safe Rooms broadly according to how they are ventilated.

- Class 1: Ventilated Safe Rooms where filtered outside air is supplied in sufficient quantity to produce significant overpressure.

- Class 2: Safe Rooms where air is filtered, but the system does not produce sufficient overpressure to provide an effective barrier to infiltration. This includes both rooms that simply have filtered recirculation of the air in the Safe Room and systems configured similarly to Class 1, but which don't provide sufficient flow rate to create overpressure.

- Class 3: Simple sealed Safe Rooms with no air filtration.

Whilst it is useful to classify Safe Rooms in this way to illustrate the different ways that they may be configured and provide a reference for discussion, designing or designating a Safe Room to meet a particular Class definition may be a rather limited approach.

The effective Class of a Safe Room may vary with operating conditions or by choice. For example, unusual atmospheric pressure conditions or unforeseen filter loading[8] may limit the overpressure possible in a Safe Room designed to operate as Class 1. Similarly a tightly sealable Safe Room may be operated as Class 3 in the initial stages of an incident to avoid filter loading and fractional pass-through whilst the external concentration is at its highest and subsequently ventilated and operated as Class 1 where necessary to maintain air quality.

It is clear that a Safe Room needs to be specified to serve a particular location, population, spectrum of threats and method of operation, rather than simply to meet a descriptive Class definition. This is the approach outlined in this Section.

The processes outlined below show the essential steps needed to determine the top-level technical requirements for an effective Safe Room for a particular situation, and to evaluate the potential suitability of candidate rooms or spaces to be converted.

At its most basic, the design of a Safe Room could simply involve selecting a room large enough to provide shelter and air for one or two hours and sealing it adequately to prevent significant infiltration during that time. This would be equivalent to pre-prepared Sheltering in Place (Sorensen J H, Vogt B M 2001). However for a Safe Room to be both effective as part of a planned response to the threat of a hazardous materials release, and also cost effective the exact nature of the threat and the physical and operational environment need to be taken into account.

The specification for an effective safe room to serve a specific building will depend on a large number of factors. There is no universally applicable specification or solution; however, there are certain analyses, judgements and decisions that will be necessary in the great majority of cases and these are set out below.

Some of the analyses detailed below depend on technical factors alone, however many may be affected by other considerations, including the organisation's assessment of the threat and tolerance for risk and cost-effectiveness.

In following the prescribed process, the key design data, assumptions and decisions made should be recorded to form the outline technical specification and concept of operations for the Safe Room. The operation of the Safe Room must be incorporated into the overall incident management processes for the building to be effective. It is also recommended that these processes are undertaken with Stakeholder involvement as far as is reasonably practicable. The

---

[8] i.e. "loading" is the quantity of material on the filter surfaces – if the filters are overloaded then they become blocked.

Annexes to PRACTICE Deliverable D5.6 "Protocols for the justification of risk from residual contamination" (Hale et al 2012) provides some useful guidance on how such involvement might be achieved.

It is recommended that expert technical advice is sought throughout this process. The initial steps in the recommended process are shown in Figure 4. Steps 1 to 3 in the Figure are discussed in Section 7.1 to 7.3 below.

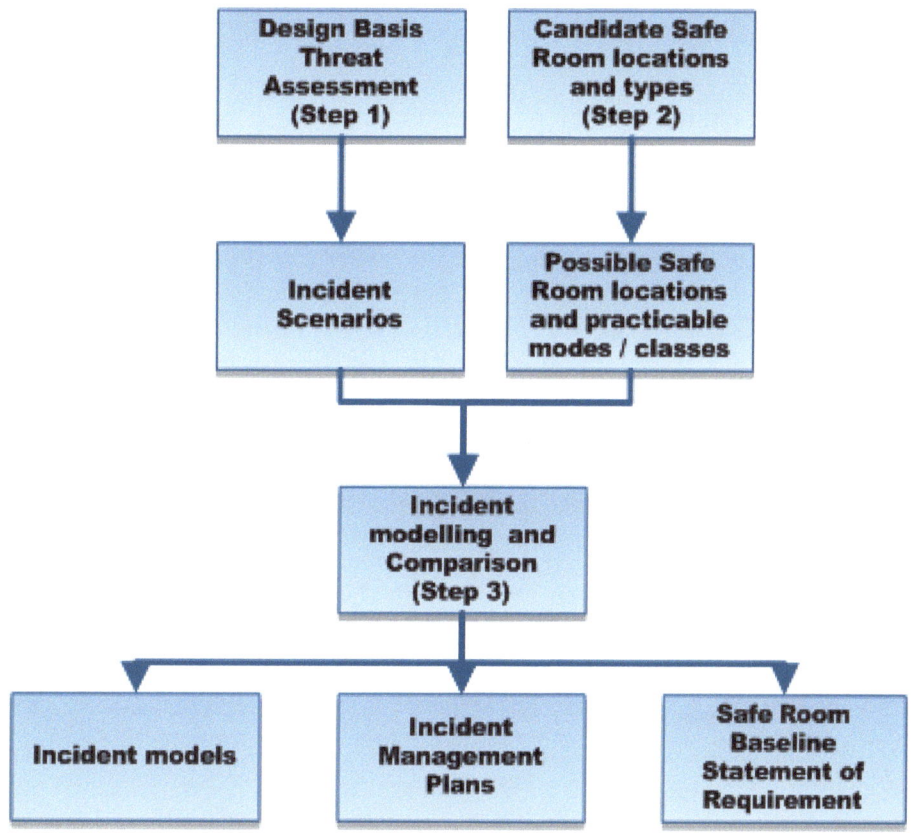

Figure 4 Development of Baseline Statement of Requirement for a Safe Room

7.1 Step 1: Design Basis Threat Assessment

Design Basis Threat is a concept that can be readily applied when determining both the requirement for and specification of a Safe Room. Essentially the concept requires that the specific class of problem to be addressed by the Safe Room is actually defined, based on an assessment of threats of malicious release and risks of accidental release. This is then used to define the Design Basis Threat. The process is discussed below.

An assessment of the capability, knowledge and motivation of potential malicious attackers, together with an assessment of the presence of and risk posed by hazardous materials, should be used to develop a range of plausible representative release scenarios that may be faced by the organisation[9],[10].

These release scenarios may typically be described in terms of

- The hazardous material(s) involved

- The incident location - specific location either inside or outside the building

- Timing – the time of day and day of week, which will affect the numbers and distribution of personnel and the behaviour of contamination plumes.

- Quantity of agent – chosen based on the threat / risk assessment

- Method of dispersal

- Whether it is an accidental release or a hostile act – which will affect the risk of secondary releases

- Whether there are single or multiple releases

- Weather conditions – which will affect the spread of agent and possibly the distribution of personnel[11]

- Expected response from Emergency Services

It may be useful to use an approach similar to that presented in FP7 Project IMCOSEC (Hale 2010) where a list of threats is developed according to a taxonomy such as that shown in Figure 5 which has been adapted from Hale 2005.

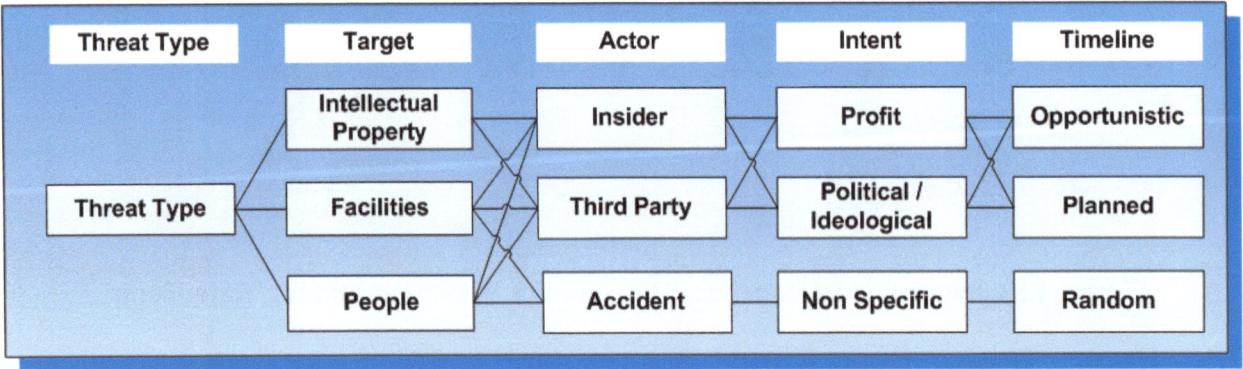

Figure 5: Example Threat Taxonomy

---

[9] Note that Design Basis Threats are not intended to cover all possible threats but those threats that the organisation wishes to design for. The selection of the design basis can therefore include considerations of probability, consequence and cost to defend against (for example). The Design Basis Threat can be seen as a statement of an organisation's risk tolerance. Advice from Security organisations and Stakeholders should be sought when identifying the Design Basis Threat(s).

[10] Users should obtain specialist support to produce vulnerability assessments and appropriate scenarios.

[11] This may be especially important in respect of agents which become volatile at temperatures within normal daily ranges for the country under consideration.

It is important that these scenarios are developed and documented in a standard format, to enable comparison and read-across between scenarios and to allow them to be interpreted consistently and used as a firm basis for decision making. A general template for developing incident scenarios and a list of some historical incidents has been produced as part of Project PRACTICE (D2.1) (Endregard M). It is recommended that this be used as a starter for the recording of the Design Basis Threats where possible.

The identified threat scenarios may then be ranked by the organisation using a simple technique such as that shown in Figure 6 (adapted from Hale 2010), where threats are placed by a group of stakeholders (by consensus) on a two dimensional chart with the axes shown. Guidance on selection of stakeholder groups can be found in Hale et al (2012). Other ranking approaches can be used but the graphical method shown is easiest for a large group to use.

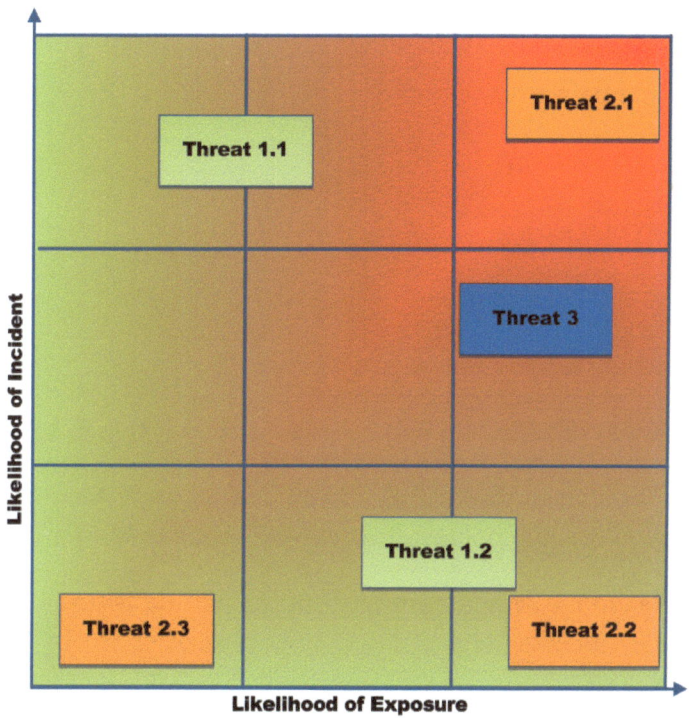

Figure 6 Example Threat Ranking Method (Different colours represent different threat types)

The nominated Stakeholder group can use the completed chart to identify the range of threats that they wish to design for – or, equivalently, to compose a single threat scenario that encompasses the selected threats.

7.2   Step 2: Identify Candidate Safe Room Locations and Types

Following the general guidance in Sections 3 to 6, and considering the building layout and occupancy requirements identify potential locations for Safe Rooms and the potential types of Safe Room that they are potentially compatible with. At this stage it is not necessary to identify all of the technical details in full but merely to identify the potential locations. These are the candidate safe rooms.

7.3 Step 3: Incident Modelling and Comparison (Develop responses to Design Basis Threats and determine the utility of Safe Rooms)

The need for a Safe Room can be tested by comparing likely outcomes (from the Design Basis Threats) with the Safe Room (assuming that the candidate Safe Room works as intended) and without the Safe Room. It is possible that in some cases, the Safe Room would not need to be used at all or it will have marginal benefit.

The specific requirements for the Safe Room can then be better defined because in principle it need only be specified to address the Design Basis Threats where it would both be used and contribute to a better outcome.

This will provide the baseline statement of requirement for the Safe Room, including:

- the number of people to be accommodated

- the potential Safe Room location

- the maximum duration of stay

- the range of agents to protect against

- whether power and other facilities will be available.

- whether it is necessary to provide ongoing operations from within the Safe Room.

This comparison should be part of the overall Safety and Security Management process and will drive a range of other requirements, not just the possible requirement for a Safe Room.

The validity of existing incident management procedures can be tested by examining what would happen for each of the scenarios defined.

7.4 Step 4: Refine the Safe Room Requirements (Figure 7)

The baseline statement of requirements, together with an analysis of costs and wider risks can be used to generate a final statement of requirement for the Safe Room. It may, for example, be considered desirable to have a Safe Room, or a more highly specified Safe Room simply to make people feel safer, or to mitigate any risk of flaws in the scenario analysis or unknown future threats. Alternatively, if particular risks are identified as key cost drivers for the Safe Room, mitigating those risks by other means, or agreeing that they can be tolerated may reduce the cost of the Safe Room.

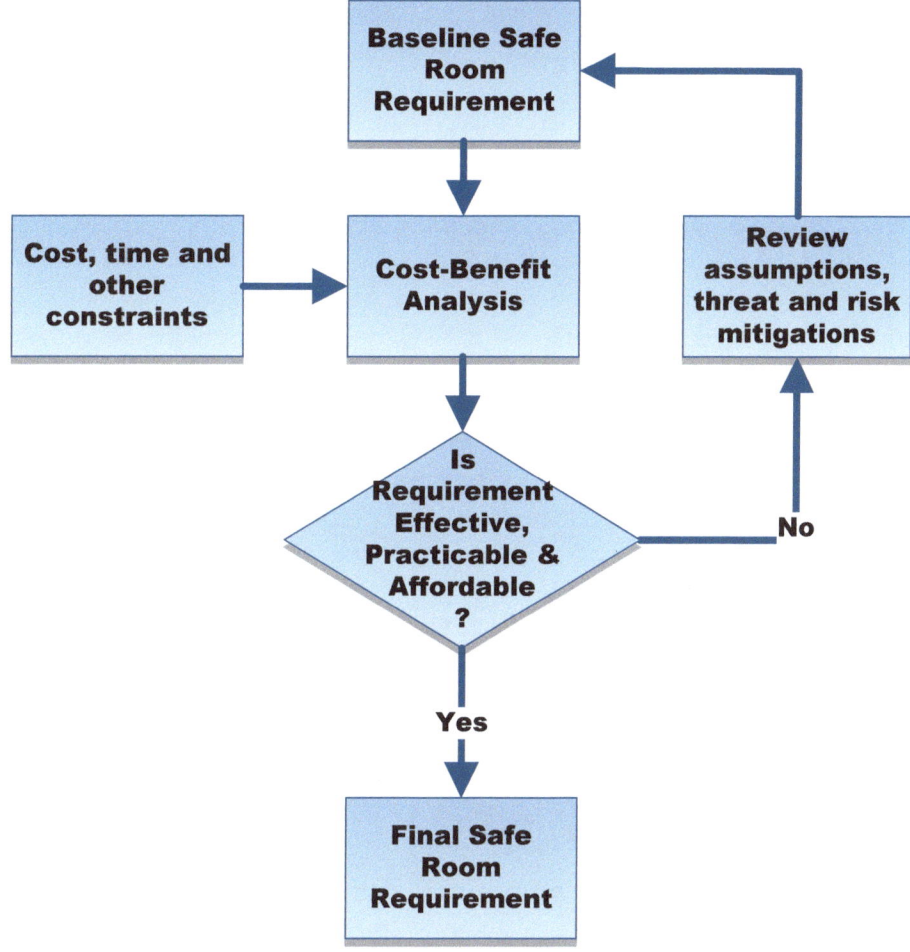

Figure 7 Refining the Safe Room Requirement (Step 4)

A fundamental part of this process is that the scenarios, assumptions, risk and threat assessments and technical data that drive the final requirement should be documented and regularly reviewed to confirm that the Safe Room and the operating procedures developed alongside it will still be effective.

It should be clear that in many cases, determining the most appropriate statement of requirement for a Safe Room will be an iterative process.

Once the requirement for the Safe Room has been confirmed and documented it must be developed into a technical specification and construction design suited to the building. The complexity of this process will be highly dependent on the characteristics of the building and the various constraints that it may impose. Retrofitting a Safe Room into an existing building is likely to be significantly more expensive and difficult than for a new construction.

If the derived requirement is for a Safe Room in a new building then Step 5 is applicable. Otherwise if it is to be retrofitted into an existing building or added onto it then Step 6 is applicable.

7.5 Step 5: Safe Room (New Build)

The specification of one or more Safe Rooms to be incorporated into a new build design is relatively straightforward. The required room size, facilities, and air supply to satisfy the statement of requirement can all be taken from existing guidance. The building ventilation systems can be designed alongside the Safe Room to make the building more resilient to CBR and other incidents, including fire, by compartmentalising the air spaces within the building and providing appropriate filtration and control.

Publicly accessible areas including lobbies, mailrooms, loading areas etc can all be designed to have separate air spaces and designed to operate at negative pressure compared to the rest of the building to control infiltration and thereby help to contain an incident at its most likely locations.

Incident management systems and policies developed as part of the building design and commissioning can incorporate the Safe Room, for example when defining escape routes the Safe Room may be treated as an "Emergency Exit" and "Assembly Point" valid for certain types of incident.

The incremental costs of designing a Safe Room into a new building may be relatively small, simply requiring certain parts of the building and systems to be configured one way rather than another.

7.6 Step 6: Safe Room (in an Existing Building – Figures 8 and 9)

The specification of a Safe Room for an existing building is more complex, because the existing structure and systems impose constraints on the possible location of the Safe Room and the facilities that can be provided. There may be several candidate locations that could meet the statement of requirement, each with separate advantages or disadvantages, constraints and associated costs.

Constraints on the Safe Room location and design imposed by the existing building will affect the cost of the Safe Room and also its applicability and effectiveness in dealing with specific incident scenarios. Each candidate Safe Room location will need to be evaluated separately. It may well turn out that several possible locations are each effective (or, more importantly, ineffective) for different incidents.

The Safe Room must satisfy the key functions of Isolation, Sustenance, Access and Communications in meeting the statement of requirement.

Once an initial survey has identified a space that appears capable of being adequately sealed and is in a sheltered part of the building, that space can be evaluated by first examining the key parameters of adequate space and air supply and then confirming that each of the other necessary facilities and functions can be addressed.

7.6.1 Step 6a

If an Unventilated Safe Room is considered as an option its feasibility should be evaluated based on the process shown in Figure 8 which first examines if the candidate Safe Rooms are large enough (given the requirements for prevention of CO_2 build-up) and then whether the room can be adequately sealed to prevent ingress of contaminants. If the conclusion is that the room cannot be sealed then the chart branches to consideration of ventilation requirements as shown on Figure 9.

7.6.2 Step 6b

If an Unventilated Safe Room is found to be not feasible, or a Ventilated Safe Room is needed to meet the Requirement, then the Ventilated Safe Room should be evaluated on a similar basis as shown in Figure 9, which after considering the same initial set of parameters then subsequently looks at ventilation and filtration requirements.

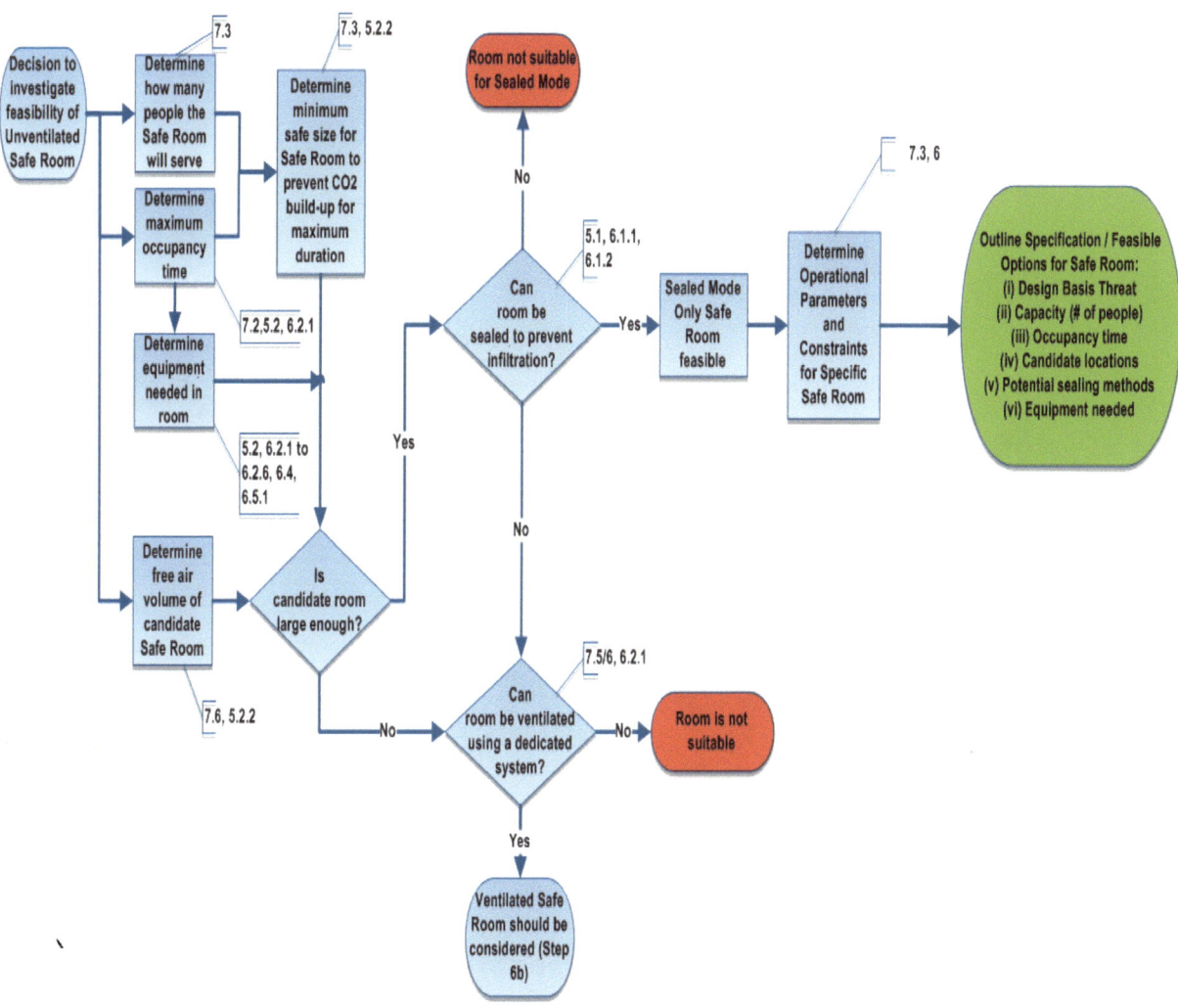

Figure 8 Feasibility of Unventilated Safe Room (Step 6a) – Callouts refer to main report sections

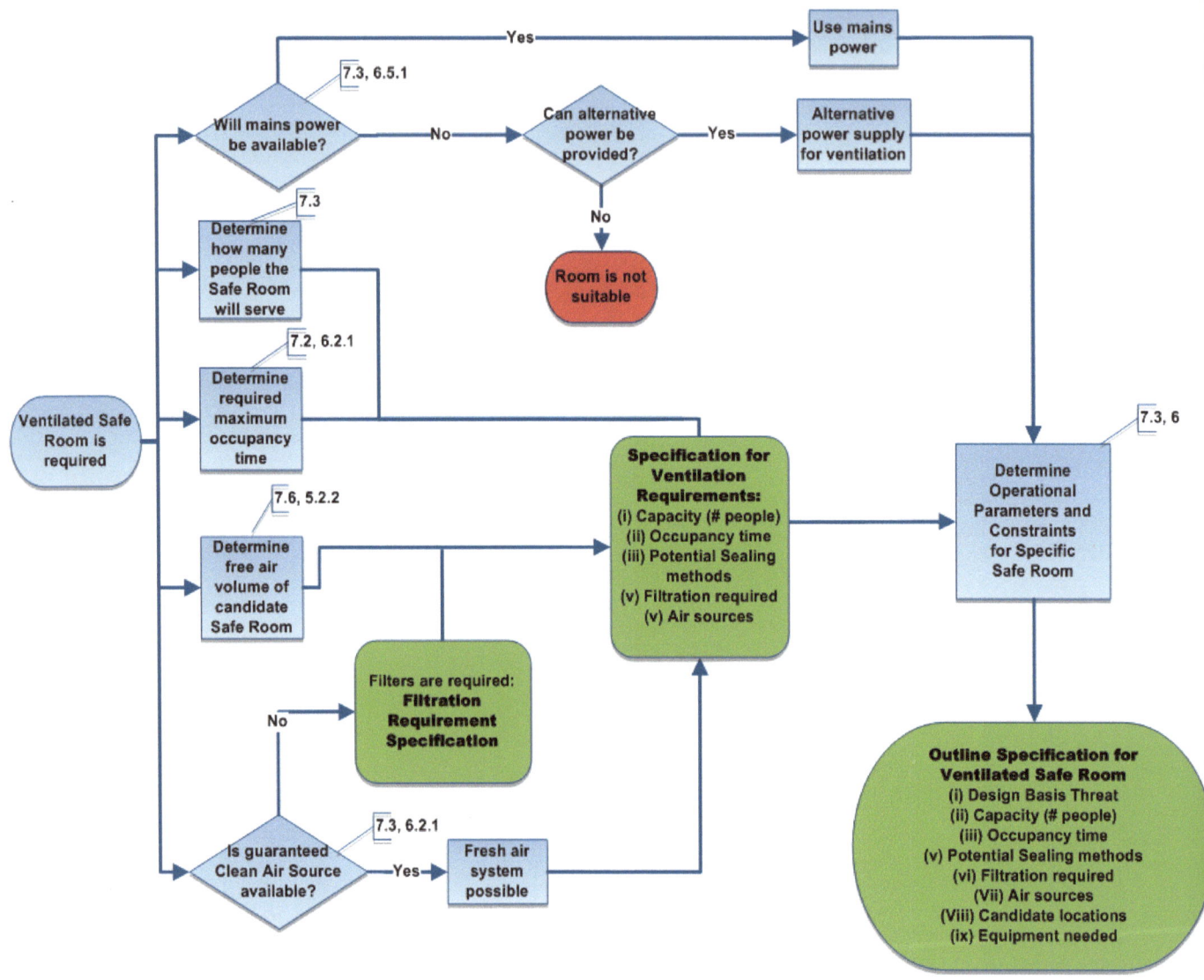

Figure 9 Feasibility of Ventilated Safe Room (Step 6b) – Callouts refer to main report sections

8. Conclusions

A high level summary of the requirements for Safe Rooms has been presented (Sections 1-4). The Functional Requirements (Section 5) can be summarised as;

- Isolation

- Sustenance

- Safe Access

- Communication

For each of these a discussion of the principal issues and available best practice guidance has been presented.

Practical difficulties, dependencies and constraints likely to be encountered are identified throughout the process and in the best practice guidance (Section 6).

A logical process of defining the requirement for a Safe Room specific to a particular organisation can be based around the concept of Design Basis Threat (Section 7). A simple methodology for the identification and ranking of these Threats has been presented, based upon earlier FP7 work.

Consideration of the issues associated with the activation of a Safe Room by commercial organisations (rather than military organisations for example) has led to the conclusion that the benefits from inclusion of decontamination and monitoring facilities are likely to be outweighed by the difficulties and potential risks presented by their operation. Furthermore, if such facilities are required prior to allowing personnel into the Safe Room then it is likely that it is already too late to use them and that in many cases evacuation, decontamination and medical treatment by the emergency services will be a more appropriate response. Thus Safe Rooms – as set out in this report – are proposed as part of a planned response to a raised threat or a response to an imminent threat rather than a response to an incident which has already caused personnel contamination. There are also ethical considerations regarding permitting some personnel access to a Safe Room but excluding others, which need to be addressed before such facilities could be implemented. Such considerations are outside of the scope of this report.

The process presented, provides an alternative approach to the simple selection of existing off the shelf products and the retention of documentation developed during the process will provide evidence of due diligence and consideration of Stakeholder input.

9. References

Blewett WK, Area VJ (1999) *Experiments In Sheltering In Place: How Filtering Affects Protection Against Sarin And Mustard Vapor,* ECBC-TR-034, Edgewood Chemical Biological Center, U.S. Army Soldier And Biological Chemical Command

CDC (Center of Disease Control and Prevention) – *Emergency Preparedness and Response – Learn How to Shelter in Place*, http://www.bt.cdc.gov/preparedness/shelter/, viewed September 2012.

CIA (2003) *Terrorist CBRN: Materials and Effects*, Central Intelligence Agency, Directorate of Intelligence, May 2003

CMHC (Canada Mortgage and Housing Corporation – 1998), *Technical Series 98-109 - Air Permeance of Building Materials,* CMHC-SCHL (available on-line at www.cmhc-schl.gc.ga/publications/en/rh-pr/tech/98109.htm)

Conway et al (2011) *Atmospheric Carbon Dioxide Dry Air Mole Fractions from the NOAA ESRL Carbon Cycle Cooperative Global Air Sampling Network, 1968-2010, Version: 2011-10-14, [Ochsenkopf, Germany (OXK) dataset]* Conway, T.J., P.M. Lang, and K.A. Masarie (2011), http://www.esrl.noaa.gov/gmd/dv/data/index.php?parameter_name=Carbon%2BDioxide&pageID=15&sortby=year&site=OXK viewed June 2012

CPNI (2010) *Protecting Against Terrorism - Third Edition*, UK Centre for the Protection of the National Infrastructure

Foster Miller Inc (October 1983), Development *of Guidelines for Rescue Chambers, Volume 1*, for US Bureau of Mines contract JO387210

FEMA 361 (2008) *Design and Construction Guidance for Community Safe Rooms* Second Edition Federal Emergency Management Agency, August 2008

FEMA 426 (2003) *Reference Manual to Mitigate Potential Terrorist Attacks Against Buildings* Federal Emergency Management Agency, December 2003

FEMA 452 (2005) *A How-To Guide to Mitigate Potential Terrorist Attacks Against Buildings,* Federal Emergency Management Agency, January 2005

FEMA 427 (2003) *Primer for Design of Commercial Buildings to Mitigate Terrorist Attacks* Federal Emergency Management Agency, December 2003

Hale N (2010) *Summary Report on Security Matrix Assessment for IMCOSEC*, Seventh Framework Project Contract SEC 242295, Deliverable 2.2.

Hale et al (2012), *PRACTICE WP5 deliverable D5.6 "Protocols for the Justification of Risk from Residual Contamination"*, Hale N and Kelly D PROJECT PRACTICE

Harper Dr P (2011) *Assessment of the major hazard potential of carbon dioxide (CO_2)* Health and Safety Executive (HSE)

HMG (2006) *Evacuation and Shelter Guidance - Non-statutory guidance to complement Emergency Preparedness and Emergency Response & Recovery,* HM Government, October 2006 ISBN:0 7115 0477 6

IAEA (2009) *Development, Use and Maintenance Of The Design Basis Threat, IAEA Nuclear Security Series 10*, STI/PUB/1386 ISBN 978–92–0–102509–8, International Atomic Energy Agency Wagramer Strasse 5 P.O. Box 100 1400 Vienna, Austria.

Kinra S et al (2005) *Evacuation decisions in a chemical air pollution incident: cross sectional survey* S Kinra, G Lewendon, R Nelder, N Herriott, R Mohan, M Hort, S Harrison, V Murray British Medical Journal, Volume 330, 25 JUNE 2005

MEFB (2011) *A Best Practice Approach to Shelter-in-Place for Victoria*, Issued by the Metropolitan Fire and Emergency Services Board, May 2011

Moran D et al (no date) *Determining Tolerance Time In Encapsulated Rooms-Physiological Considerations,* D. Moran, A. Cohen, S. Ashkenazy, D. Albukrek, I. Tur-Kaspa. Institute of Military Physiology,NBC Medical Branch ,IDF Medical Corps, Israel

Morrison RW (2001) *NBC Filter Performance (ECBC-TR-135)* Edgewood Chemical Biological Center, Research And Technology Directorate, U.S. Army Soldier And Biological Chemical Command, October 2001

NIOSH (2003) *Guidance for Filtration and Air-Cleaning Systems to Protect Building Environments from Airborne Chemical, Biological, or Radiological Attacks*, NIOSH 2003-136

NIOSH (2007) *NIOSH Pocket Guide to Chemical Hazards*, Department Of Health And Human Services, Centers for Disease Control and Prevention, National Institute for Occupational Safety and Health September 2007, DHHS (NIOSH) Publication No. 2005-149

NIOSH/Foster-Miller (2007) *Chapter 4, Atmosphere Management from Foster-Miller Phase II Draft Final Report "Refuge Alternatives in Underground Coal Mines,"* NIOSH Docket: 125 Dec. 2007

Endregard M et al (2012), *PRACTICE WP2 Deliverable D2.2 "Scenario template, existing CBRN Scenarios and Historical Incidents,* PROJECT PRACTICE.

Shea D A (2003) *High-Threat Chemical Agents: Characteristics, Effects, and Policy Implications* Congressional Research Service RL31861 Updated September 9, 2003

Sorensen J H, Vogt B M (2001), *Will Duct Tape and Plastic Really Work? Issues Related To Expedient Shelter-In-Place* ORNL/TM-2001/154 Oak Ridge National Laboratory for Federal Emergency Management Agency, August 2001

US-Army Corps of Engineers (1999) *Design Of Collective Protection Shelters To Resist Chemical, Biological, And Radiological (CBR) Agents,* Technical Letter No. 1110-3-498 24 U.S. , February 1999

WHO (2005) *Minimum water quantity needed for domestic use in emergencies* World Health Organization, WHO – Technical Notes for Emergencies, Technical Note No. 9, January 2005

Yantosik G (2006) Shelter-in-Place Protective Action Guide Book ANL/DIS-06/25

Further information:

FEMA 453 (2006) *Design Guidance for Shelters and Safe Rooms,*
Federal Emergency Management Agency, May 2006

Lindstom, G (2004) *Protecting the European homeland - The CBR dimension*, Chaillot Paper No69, Instutue for Security Studies, European Union, Paris July 2004 ISSN 1017-7566

CIA (1998) *Chemical/Biological/Radiological Incident Handbook (October 1998)*
https://www.cia.gov/library/reports/general-reports-1/cbr_handbook/cbrbook.htm Viewed Jun 2012

Cousins D, Campbell SD (2007) *Protecting Buildings against Airborne Contamination* LINCOLN LABORATORY JOURNAL, VOLUME 17, NUMBER 1, 2007

Persily A, (2004) *Building Ventilation And Pressurization As a Security Tool,*, ASHRAE Journal ashrae.org September 2004

CSEPP, *US Chemical Stockpile Emergency Preparedness Program (CSEPP), Available at* http://www.cma.army.mil/csepp.aspx, *May 2006*

1. Annex1: Typical Safe Room Checklist

Check the following for number and condition:

- Chemical Toilets and supplies

- Furniture

- Food

- Water, cups

- Lighting / Batteries

- Sealing materials (Duct Tape etc)

- First Aid kits

- Sanitary supplies and equipment

- Safety knives (for mass emergency exit from plastic sheeted safe rooms)

Functional Check/Inspect the following

- Communications equipment

- TV, radio, internet

- CO_2 and O_2 monitors

- Fire fighting equipment

- Local ventilation equipment (e.g. desk fans)

Check Availability of the following:

- Access and egress routes still viable

- Entertainment materials (games, cards, mp3 player etc)

- Writing materials

- Contact numbers and e-mail addresses etc

- Personnel List (Muster) logs etc

- Washing materials

- Emergency Plans, operating manuals, procedures etc for equipment

- Threat Assessments and Procedures

- Signs to indicate that the room is activated (sign outside, red light etc)

- Occupancy v Time v CO_2 Charts

Inspection / Check / Maintain the following:

- Air tightness – check seal condition etc

- Heating and Ventilation checks - Air intakes (unblocked, protected etc), filter condition

- Changes to the local environment etc which may change hazard potentials etc

- Spare sealing materials (duct tape etc)

- Thermometer

- Manometer (for monitoring air pressure differentials in ventilated rooms)